# Mysteries:

# From Darkness to Light

# – Hidden to Sight

Thoughts Concerning Home Schooling

and

Hosting the Idaho Science Olympiad

1993- 1994- 1995

By Bonnie Meyer

Chapters:

## Mysteries:

It has been said; that everyone needs a great purpose for their lives, to be a part of a great love story, and to be involved in a great mystery.

A mystery is a profound secret. Something unknown or something kept cautiously concealed and therefore it is not comprehended until it is explained, then it causes us to wonder, and become excited as a result. In the Bible Job said that "God reveals mysteries and takes from darkness and brings deep darkness into light". In Corinthians it tells us if we are servants of God he will reveal a mystery to us, and then we are to be stewards of that mystery. When Jesus walked on the earth he spoke in parables or mysteries to illustrate the mysteries of walking in the Kingdom of God, or the mystery of Christ and His wisdom is said to be in the form of a mystery.

A secret is something that is concealed from notice. Secrets are usually private and secluded. They are removed from sight or privately known but not apparent in the operations of everyday life. There can be secrets in the physical world, secrets in the spiritual, things hidden from us mentally and socially. Secrets can affect our lives if we recognize the factor not.

This book is or can be both a mystery and a secret. It deals with our lives and the 7 mountains that we are involved with daily, even if we don't recognize this fact or even if we have never heard of the 7 mountains. It deals with secrets, the reactions of Science and the reactions of people. The things that we perceive become no longer a mystery to us but we can use that gained knowledge to affect and better our lives, with the understanding of ourselves and our responses. It can help us encourage others on their paths, because each of us deal basically with the same issues of life. The mystery is how we respond to those encounters and mountains.

No matter what our age, we each have the opportunity to be the next Sherlock Holmes, Nancy Drew, Encyclopedia Brown, Theodore Boone, or MacGyver. It does not matter what gender we are, male or female. It does not matter our education, family status, religion or anything outside of ourselves. It simply does not matter, but it does involve us, our gifts talents and our abilities to make choices. It involves our ability to see and observe our surroundings and respond correctly to each given situation. Yes, to even put ourselves in harm's way to help or defend others

I think the greatest purpose was accomplished by the Creator God. Who formed and fashioned the world in which we live, with amazing colors, resources from the earth, air to breathe, water to refresh ourselves, and friendships and fellowship that help us to thrive. Each of us has been given the ability to choose and the ability to use our five senses over our own lives and the circumstances that we find ourselves in making decisions, by observation and sound judgment. Whether it be our relationships, our health or our lives in any way that affects us or those we associate with. It is amazing to think He created this wonderful world in 7 days.

The greatest love story, I believe, was how Creator God sent forth His own son in the form of a man, Jesus of Nazareth to win back His creation that disregarded His instructions and listened to an enemy voice that was contrary to life, and to the cautions given at creation. This caused the consequences that brought destruction to relationships, and to the garden which the first couple was given to care for and have dominion in. We may come again into fellowship with our Creator to know His plans and purposes of why He created us, and redeemed us from a world that chose not to include Him. At times life is not easy, but it is possible to admit our need. Until we come to Him, it is very easy to admit we are missing the mark without Him; but to repent and ask Him to come into our lives and give us a heart to know Him and a teachable attitude to learn from Him, is a bit more challenging because of our insecurities and unbelief. We come to our Creator through the Lord Jesus to the light and recognize the sacrifice He gave for us by making a way for us to return to the Father's original intent and for His creations around us. For those of you who do not believe in Jesus, that He was a real person and did what it is said He did. Just type in the date 2016 –We just proclaimed His life, death and resurrection until He comes again. For our Jewish friends it is 5777, which tells us that there is still Grace available to us, as we decide how we want to live our lives and what battles we need to be involved in or what struggles we will face. Five is the number of grace in the Hebrew language, and seven is a picture of a sword, or provision, or completion. Jesus came in grace to defeat an enemy we sometimes cannot recognize to provide for us health and resources in His name that we might live life to the fullest.

The Greatest Mystery for me is why God has created me, redeemed me, and sent His Holy Spirit to draw me closer to know Him giving me specific gifts and abilities. Now because of that, I am able to be a light for others to come to know Him and His amazing love for us, and bring a blessing to His Kingdom with the gifts, talents and abilities He has given me, and impact the world around me by my walk with Him and the wisdom and choices available to me. This is one reason 5777, reminds me of 7 days of Creation, the 7 hours Jesus hung on the cross, and for the 7 lights of Menorah that represent to me the 7 Spirits of God and the 7 gifts of the Holy Spirit.

To begin my journey with this book, I have given you some background information, but not the reason for writing and putting it into a written form. I chose to homeschool my 4 children, one who was interested in Science, and so I began to fulfill his desire to learn different areas of the sciences. In order to teach him, I had to first teach myself and so I sought out a group of like-minded parents and families, and we chose to be a part of Science Olympiad.

Science Olympiad teaches science skills, but also has competitions where schools are able to come together and compete against each other for the chance to go to the National Science Olympiad. I was asked to be in charge of the Crime Buster Event, hosted at the College of Southern Idaho in Twin Falls.

Crime Busters is where a student uses fingerprinting, ink analysis, and chemistry to determine liquids, solids and metals, and then those skills are used to solve a crime scenario. I was given the task to somehow make a test that would accomplish this. I called my local Police station and 3 Detectives were excited to be a part of the event. They helped me to put together the first test and they provided the finger prints. Off we went in trying to put together a scenario where the students could test and accomplish the goals of the perimeters given for me to work with. These tests are not copyrighted because I wrote them with the help of the officers, and they have remained in my file cabinet for 20 years, but I thought they might benefit others at this time.

In putting together this book from the actual event, I have included the test answer keys, and attempted to add the fingerprinting cards to no avail. Police documents are not the regular size so cannot be duplicated for obvious reasons. I had removed the names of the detectives that helped me as they have since retired from the force, but the cards were not easily incorporated into the text, even though I had permission to include them so I did not use them. Since fingerprints are easily obtained I will let you provide your own fingerprints, and shoeprints if you decide to reenact the events.

In the chemistry side of the testing, I have included those tests as well but not any specifics on how to test them except as the general information as I have listed. If you do, be extremely careful. I am not liable for any of your decisions to do so, or the problems you encounter. If you need help please contact your chemistry department at a school, or buy a book to teach you the proper procedures and safety instructions.

To start off I would like to bring to light the fact that each of us have what is called 7 mountains in our lives. Each student and teacher I encountered, each family no matter who they were or where they lived, could be identified in how they used those mountains in their lives. Some were wealthy, and some were not, and some were on a strict budget so that I provided the chemicals and equipment. Each of us function in and impact each mountain by our leanings and gifting. Each mountain does at times overlap, but they are outside gender, economic status and personal preferences. We are able to choose and move into different areas as we desire. At different times in our lives we may change from a profession to try out another mountain of influence.

In hosting the Science Olympiad Crime Buster test I dealt with the Family Mountain; Education Mountain of public school, private school and home schools, and the College of Southern Idaho for any resources I needed during the event. The Business Mountain were those businesses that donated the chemicals and equipment. The Government

Mountain was in the form of the Police officers and Detectives to help the students see the role of Police Officers in the lives of our communities, homes, schools and businesses and how they, the students themselves can be a part of any solution to any problem or crime by their own observation skills. You can find the resources to study the 7 mountains more in-depth if you desire to do so.

## My 7 Mountains:

Religion Mountain –what we believe or who we believe in God, humanism, Satanism, anything that we worship and adore giving our time and resources to. For me personally, religion is man trying to reach God whereas Christianity is God coming down and reaching man by His initiation not mans through the person of the Lord Jesus Christ.

In religion, there are 3 ways that religious groups try get you to believe what they do: 1. By a Barbaric Mindset in which if you do not agree with them, they use violence against you. Hence throughout history and today we can see that mindset in action by violence and destruction. 2. By a Greek mindset. The Greeks valued knowledge, arguments and logic. So someone who functions in that mindset will try to talk you into what they believe, even using shame and ridicule if it is to their advantage, wearing you down to finally agree with them and submitting to them even if you do not. 3. Is a Hebrew Mindset. It is an individual Mindset where the individual is given free will to make decisions and decide for themselves, without pressure, but they are held accountable to a power higher than themselves. The Constitution of the United States was modeled after this mindset. If you desire, you can read the Bill of Rights.

Family Mountain—each of us has a family, or someone, or some group we feel a part of even if it is an extended group and not a biological family as such. It is a place or should be a place, where we have food and shelter, and hopefully gain friendship and acceptance. We can have a pride in our family name or embarrassment of it. I believe our name is the only thing we keep throughout our lives, and so we should protect and value our name and our reputation which comes accordingly with it. Our families can narrow by death or divorce, or expand by marriage or adoption. We can determine how large we desire our family group to be by our own love and concern for others, or how small by self-protection. Even within the other mountains of influence we are able to interact with like-minded people; where we worship, where we shop, or the hairdressers we trust etc. We can include them and celebrate their lives and milestones with them and allow them access to celebrate ours.

Currently, society is obsessed with genealogy, of blood lines and where you came from. I would encourage you to realize you can have blessings from our ancestors, but also difficulties inherited, but we do not have to have the traits or the diseases that were in our blood lines. The sayings —" Mother like daughter", or "Father like son" do not have to be true, or believed; we are each individuals, even though we may get our hair color, or eye color from our parents it does not mean that we inherit their traits or personalities as our own.

When we come to the Father through the Lord Jesus, we become a part of His family. He says that He adopts us into His family, and then we have rights and privileges as a son or daughter, and also an inheritance granted to us by His divine authority.

Education Mountain–we all have the ability to learn and adapt to our environment. Most of us have endured public schools, private schools, trade schools, colleges or universities. Each of us have the ability and the opportunities to learn something new each and every day. Our gifts and talents may then lead us to continue in that mountain in the form of becoming Teachers, or Homeschooling as I did. Or being a part of child care profession, caring for the next generation by imparting the things we know and the manners and traditions we have learned. We do not have to be labeled smart, stupid, or inept. We have abundant resources in books, libraries, and online resources to gain the knowledge needed in any subject we are interested in. We can then use that knowledge gained to impact those other areas of our lives and to those around us.

In the Bible, the Lord said that "He would be our teacher". "Call upon Him and He will teach us great and mighty things we do not know." All we have to do is ask. In the book of Jeremiah, he was told that the Lord was the creator of all flesh and nothing is too difficult for Him. All we have to do is come before Him and ask. Maybe we feel God is silent and not speaking to us, but are we speaking to Him, by coming and asking Him daily for His input. In the New Testament in the book of James, it says, "if anyone lacks wisdom let him ask of God who gives generously to all men". Do you have an issue go to the Lord God and simply ask; He is waiting for your questions.

Government Mountain –families, cities, states, nations –they bring order into our lives. Hopefully warding off lawlessness so we may live peaceful lives. We may choose be fire fighters, police officers, Politicians, judges, lawyers, someone desiring to help our surroundings become a safer environment, or to help and rescue those in need. We all need to respect those officers of the law and bless them, instead of disregarding them and mistreating them with our words and actions.  When we see them, we need to recognize they are there to help us, and at times they are our first line of defense for ourselves and families, homes and property.

In the Kingdom of Heaven it is a benevolent dictatorship. The Lord Jesus is King, and we are part of His Kingdom, representing Him on earth.  In the Lord's Prayer, He said to pray for the order of heaven to be brought to earth. There is no sickness, poverty or lack in heaven.  God has the resources to release them you if you submit to His Lordship.

Arts and Entertainment Mountain –what is forming your values? The books you read or the things you watch? The people you have fun with, the sports or competitions you are involved in?  Do you like to act, sing, and dance?  This is one reason why so many try out for America's Got Talent or other shows that help expose their talents and gifting.  Some are successful and some are not, but they chose to try.  We might pursue to write our thoughts in the form of books to impart our wisdom, or to entertain others by the genre we choose. Each of us can learn to recognize reality from non-realistic situations.  There is a difference between the actor and the role they play, though some have had difficulty in separating the role and action with reality.  Each of us does something to relax and be entertained even when nobody's watching.

In the Heavenly standard, you do not have to wear a certain style, have your hair a certain color, wear a certain brand.  Come as He created you and you are welcome.  He loves from the inside out. The Arts and Entertainment Mountain looks on the outside, but God says He looks at the heart.

Business Mountain – are we employed or we are self-employed.  It has been said we work so we can provide ourselves food and clothing and shelter, so we can get a job to work to buy food and clothing, etc.  We are able to get our self-worth or self-esteem from the things we accomplish benefiting our families and communities.  Most businesses identify a need in some area of life and choose to fill that need by providing goods or services to their communities. Business are usually expanded by word of mouth, a good reputation and timely service.

Finances Mountain –we each have a part and a say in our financial standing. We may be savers or consumers. If we are bent towards numbers, we may become contractors in construction, or accountants, financial advisers, or the dreaded tax agents that strike fear in some. We can be rich or poor; depending on our own personal choices and not relying on government handouts or ill-gotten gains acquired by illegal means. We can act with generosity or be consumed with greed for more, not enjoying what we have acquired. Will Rogers once said, "People send money they don't have, to buy things they don't need, to impress the people they don't like".  If we have "it we have to use it or protect it", as the law of entropy and decay will always be present, but in the end we will all leave our wealth and possessions for others to use or fight over.

In the Government Mountain, they are always desiring revenue in the form of taxes to promote their goods and services, or to exert control in some area of our lives. God says, give Him the first. First of time, and first of money in the forms of tithe, gifts and offerings, and He will bless you abundantly. He does not take anything from you, but as you give to Him, He blesses in return.

We not only have these 7 mountains at our disposal but we also have 5 senses to help us navigate our lives. The senses help us to discern danger or safety and acceptance; fear –flight or fight; joy and comfort; shame or blame. We can discern emotions of others and our surroundings by the use of our senses. By what we are sensing, or the expressions on people's faces or even their body language. We can use our senses to hinder others or to help others to cope with the stresses of life.

In the next chapter, I briefly remind us what our senses are, and how they benefit our lives. The senses played a part in the Science Olympiad training, and in the scenario and chemical analysis.

## Senses:

Sight –the ability to see gives us direction, and we can be aware of our surroundings. We are said to be near sighted, far sighted, short sighted, or can have hindsight or foresight. Our eyes are said to be the windows of our souls and give others a clue to our innermost being, without being aware of what our eyes are signaling to others or mirroring. We need to be cautious because what we perceive is not always true, so we need discernment and not go by outward appearances or the things presented to us by sight.

Hearing –we have the ability to hear sounds, high pitched or low pitched; sounds that are close to us, or far from us. Our ears alert us to danger or bring us sounds of comfort, or to even lull us to sleep.  To be able to communicate we must be able to hear, and be able to use our own voices to communicate.  It is said that low bass sounds can cause internal problems with our major organs, and to stir up anger. This is why in some high crime areas, classical music is played outside business as a calming factor on the streets of those communities.  We need to use caution in hearing, because not all we hear is true.

Smell –is carried by the air around us in the air we breathe. We cannot observe smells unless we notice a skunk in the road, or a pan burning our meal on the stove. We can become used to smells and odors, even unpleasant ones, or we can be drawn to aromas that appeal to us through aromatherapy. Smells can alert us to danger as to a fire and smoke, or to deceitfully draw us to things that harm our minds and bodies.  Use caution with odors, it causes allergic reactions in some people with sensitivities to chemical odors.

Taste –is the ability to judge between sweet, salty, sour or bitter.  Smells can affect our taste buds and appetites.

We have all heard expressions "it left a bitter taste in our mouth", or "that was sweet". We can taste some aromas around us, and cause us to avoid those situations, or be drawn to them. Have you ever bought a cinnamon roll because you walked by a stand in the mall and it was a temptation you couldn't resist? It is said, that in some conditions the loss of taste is said to be a first warning signal to a physical problem that needs to be addressed.

Touch –our skin is the largest organ, it alerts us to gentle caresses, pressure, or hot and cold sensations to keep us healthy. It is our first line of defense in the world around us. We can sense with our skin, as in the expressions, "it made my skin crawl" or "it gave me goose bumps". Our skin shivers to help us heat up our inner core, and sweat glands help us cool down to maintain a correct body temperature. There are right touches and wrong touches and we, by our moral code need to recognize the difference between the two. We do not need to hug someone just because they desire to hug us.

In the testing of chemicals and odors and even with sounds you must protect your eyes, ears, and yes, even your noses at times because certain substances can create gases that are harmful to lungs or throat. This is why safety goggles or other equipment is recommended; for our protection and not just because someone can demand you wear them even if you do not like the fashion statement. It is asked and required for a reason.

## Other Defining Characteristics:

Blood —our life is in our blood as the Bible tells us. It is our defense internally against infirmity and diseases. Our blood is reflected in our diets, and there are different types of blood. We can give blood, and receive our blood type from others. There are rare blood types and common blood types, but it is always red. It does not matter our skin color. So many give blood or have received blood from generous donors, and then curse someone of another skin color by prejudice and in that action are they cursing themselves. My observation is that a curse without a cause will not lite on the person it was sent to, but will go back to the one who spoke it. Maybe we need to control our lips in speaking and the words we say might not so readily come back to us. We all have said "I am catching a cold" —and caught it. Or said other foolish words that have come back to have us experience them ourselves. Maybe the prejudice is coming back on us, because it came from us, and is only returning to us.

It is said that in quantum physics you can put blood on a lie detector test, and have the person drive 50 miles away, answer the questions aloud so the blood can hear the question and the blood will not lie. It gives us something to think about anyway. In the Crime Buster tests, there may be broken windows, or scratches, but blood was not involved. There are tests you can order if you desire to test your blood to find out your own blood type. Blood types know no economic structure, or gender, but there are some religious groups which do not accept blood transfusions, and so they give blood to blood banks for their own selves if they know a surgery is needed or approaching.

In the spiritual realm, we rely on the blood of the Lord Jesus to cleanse us from all sin. We, according to the Biblical account are able to claim His blood, His Name and His

Love for any area of our lives, casting all our cares on Him because He cares for us. Sin is just missing the mark because of lack of skill, lack of focus or lack of understanding, and we all fit into that category.

Fingerprints— each person has a set of unique fingerprints. They are your own signature for the world to see. In the Crime Busters event, we did have fingerprint analysis. Within fingerprints there are: 1. Loops, 2. Whorls, 3. Arches. We provided fingerprint cards of the suspects and then copied a part of one or more of the prints to fit the scenario that we had written in the crime scene and labeled the prints accordingly. There are ways to have your fingerprints on file for yourselves as a safety precaution, and many parents choose to have cards made of their children's prints for safety purposes. Usually a police department will provide this service. Then the parents can keep those prints on file in their own homes in case of an emergency. Any other identifying traits can be added to the prints for identification purposes.

Foot or shoe prints –we did provided shoeprint samples for the scenario the third year. We took shoe prints, and laminated them, labeling them as the suspect's shoes, and the place where the print was found. Footprints and stride of shoeprints are helpful in determining the height of a person as to their walk, or any defining of limping or being hurt, such as having a sprained ankle or twisting a knee.

Ink or pen—when we test ink it is easy to put a line of ink on a coffee filter, or piece of paper, and angle the paper so that it rests in a small amount of water in a small dish or a cup. The paper will soak up the water to the ink, and as the water hits the ink, the ink will separate into colors that will define which pen was used.

In each event we chose a specific time, place, circumstances that would explain the actions or the people involved; we did not go into consequences until the 3rd competition when the students felt there needed to be some sort of moral lesson included.

**Caution: If any child or student is reading this do not test anything without your parent's' or guardian's permission. This is not a game to play, but a study of the reactions of those various powders, liquids and solids. It is essential that any study using testing you are very careful! Not only for your own safety, but the safety of others and the property around you.**

## Chemical Analysis:

We used metals, solids, and liquids. Some were common and easily obtained and others we obtained from chemical stores or had donated to us by the generosity of those businesses. The only substances that would be more difficult for you to test would be the first 3 or 4 on the list. Or you could contact your school's chemistry department. We made available to the students the glass plates to test on, the Litmus papers to test to see if it was a base, acid or neutral; PH paper. HCL –hydrochloric acid and iodine and water were all put into small bottles that had eye droppers for controlled safety amounts. The students were all required to wear aprons to protect their clothing, and safety goggles to protect their eyes. We had running water available if any of the HLC was spilled in any way or it they got it on their skin. We required them to wash it off immediately or clean it up quickly as it will eat holes in clothing, floors, and burn skin. We provided paper and pencils for each student and they usually worked in teams. Some students chose to separate the parts of the test leaving the harder questions for the last to work on together. During the event we had to have a specific time allotted for testing. There was no tasting allowed, period. If anyone was noticed tasting anything they were immediately disqualified —not because we were mean and controlling but common sense and safety ruled the day.

Copper II Nitrate —it is a bright blue color and so very easily eliminated in the testing. (Most blue food is poisonous - except blueberries for example. The usual rule is if it is blue don't eat it.)

Aluminum Hydroxide —it changes color when iodine is added but then returns to white.

Sodium Acetate —it is an irregular shape but has a PH of 8 tested on litmus paper

Sulfur—is a bright yellow color. It does cause a poisonous gas, and so is used with caution. It burns hot. (Think Sodom and Gomorrah)

Plaster —is easily obtained as it is used in construction to cover sheet rock seams being plastered out to have a smooth surface to paint. Plaster is used to cover flaws in general construction. It is used in arts and crafts and ceramics. (Do we whitewash over issues? Are we easily broken?)

Flour –is easy to obtain, at any store. Bleached or Unbleached

Sugar –is easy to obtain and is irregular shaped, and dissolves easily in water. (We need to be cautious with flattery or excessive praise.)

Salt –easily obtained. Salt is in the form of cubes. Can be easily seem under a microscope or magnifying glass. (We are said to be the salt of the earth, it causes thirst and preserves food.)

Calcium Carbonate –can grind up tablets from a health food store. It reacts to acids. Usually in found in a fine powder form. (Think bones and teeth and strength)

Sand –are small particles of different sizes and some different colors. It is gritty and does not crumble easily. (We are told to have our lives built solidly and not shifting easily.)

Cornstarch—clumps together in water, and is easily molded. When it clumps together it is harder to separate, but forms a paste very slowly. It can dry out and the clumps separate again.

Baking Soda –reacts to acids (all of us can help others rise, or be puffed up ourselves or react violently to certain situations)

Water – (liquid) clear, no smell, colorless, defies properties of gravity with surface tension

Rubbing Alcohol – (liquid) medicinal smell; antiseptic (kills germs)

Household Ammonia – (liquid) strong smell; strongly disagreeable, colorless but a pungent gas, used as a cleaner

Lemon Juice— (liquid) smell, yellow color

Zinc – (metal) Reacts to acid, has a dark appearance.

Aluminum – (metal) easily crushed, non-magnetic properties, delayed reaction to acids, silvery appearance.

Tin – (metal) less reactive to acids, used in food containers for food preservation, used to coat other metals to prevent corrosion.

Iron – (metal) dark colored and is drawn to a magnetic force. (We can be drawn by positive or negative forces.)

We encouraged the students to use a glass slide to test the metals instead of one by one. If they put a metal in each corner of a slide they could easily pass a magnet under the glass, since only one would react. Some listened and some were just raring to go, and needed not tips or counsel.

So here are the tests I prepared for the Idaho Science Olympiad Event. They are in order of the test, and the answer keys according to the 3 years I acted as the advisor for the event.

The next year the Idaho Science Olympiad had become so popular that it was moved to a larger community utilizing other college and university campuses for the competition, for a larger participation. I had enjoyed putting together the event for those 3 years, and I even now encounter 'students' of the past that enjoyed the event.

I have enjoyed putting these together not only for the competition, but for you to enjoy. I hope this book has helped you understand yourself, your choices and the world around you. If you do decide to re-do the tests yourselves. Please be careful, especially with the chemical side. The Lord has established certain reactions and they are set, and will respond the same each time tested.

# Event Test Number #1

This test was our first test we prepared together; myself and the help of the Detectives who joined me in putting together the fingerprints. We had a fun time preparing. We chose a setting to reflect individual lives. You can identify the mountains affected in the act if you desire to do so.

Start Time: _____      End Time:
_____

School Name: _____
Team number: _____
 Detective Names: _____

On April 16, 1994 at 2:00 A.M. a house at 123 4th Ave. (Town and State/Country) was burglarized.

Suspect #1. Gained entry by climbing through the window on the North side of the house. While climbing in the window suspect #1 caught and tore his/her shirt on the window. Suspect #1 went to the front door located on the South side of 123 4th Ave. and let suspects #2 and #3 into the house. Suspect #1 left fingerprints on the window and on the inside door knob.

Suspect #2. Came into the house leaving footprints on the carpet. Suspect #2 went into the kitchen and took the microwave off the counter and knocking the salt over. Getting salt on the floor. Suspect #2 took the microwave out to the waiting car.

Suspect #3. Went into the master bedroom and took the jewelry box off the back of the bed.  Suspect #3 took the jewelry box out to the car.

As suspects #1, #2, and #3 came out of the house to get into the car, the gardener came around the edge of the house.  The gardener ran after the three suspects and got to the car just as the police arrived.

The police arrested all 4 subjects.  All 4 of the subjects showed identification to the police.  The owner of the house was not home and they were not able to verify that who actually the gardener was.

Subjects were identified as listed below:

1. Joe R. Bad
2. Patty Larceny
3. I. M. Nogood
4. I. B. Bad

During the search of the vehicle a note was found with address to 123 4th Ave. and a list of Items to be taken.

The Police collected the following for evidence:

#1. Material from window.

#2. Fingerprints from window.

#3. Fingerprints from inside door.

#4. White powder material from shoe prints on carpet.

#5. Fingerprints from jewelry box.

#6. Clothing from all 4 subjects, including shoes, pants, and shirts.

#7. Fingerprints from all 4 subjects.

#8. Vehicle registered to Joe R. Bad of 156 6th Ave. West, (Town and State)

#9. 4 pens, one from each subject to compare ink with note found in vehicle.

#10. Fingerprints from microwave.


Essay and Analysis:

Please answer the following questions to solve the crime mentioned, the chemical analysis, and the essay to explain your theory and opinion of the actions of those involved. Do the chemical analysis in a safely manner, using the safety equipment at all times.

Questions: 1993 Test

1: Does material match cloth from the shirt of subject

        1._____ 2._____3._____or 4._____?

2: Who is suspect #1?

_____

3: What is the white powder suspect #2 left on the floor?

_____

4: Who is suspect #2?

_____

5: Who is suspect # 3?

_____

6: Who is the gardener?

_____

7: Who do latent prints 1, 2, and 3 belong to?

_____

8: Which fingers do prints 1, 2, and 3 belong to?

      #1 _____ #2 _____ #3 _____

9. Who do latent prints 4, 5, and 6 belong to?

_____

10 Which fingers do prints 4, 5, and 6 belong to?

      #4 _____ #5 _____ #6 _____

11: Who do latent prints 7, 8, and 9 belong to?

_____

12: Which fingers do prints 7, 8, and 9 belong to?

      #7_____ #8 _____ #9 _____

13: Which pen wrote the note found in the vehicle?

1 _____ 2 _____ 3 _____ 4 _____

Chemical List: (Each of the solids, metals and liquids were placed in small containers for easy testing.)

Identify the contents of vials 1 through 17, Two (2) of the vials contain mixtures of two (2) substances.

1. Copper II Nitrate          _____

2. Sodium Acetate          _____

3. Aluminum Hydroxide          _____

4. Aluminum          _____

5. Flour/Plaster          _____

6. Tin          _____

7. Iron          _____

8. Sugar/Salt          _____

9. Sulfur          _____

10. Calcium Carbonate          _____

11. Sand          _____

12. Lemon Juice          _____

13. Household Ammonia          _____

14. Rubbing Alcohol          _____

15. Water          _____

16. Zinc          _____

17. Cornstarch          _____

1: Does material match cloth from the shirt of subject?

   1._____ 2._____3. __X___ 4._____

2: Who is suspect #1?

   _____Patty Larceny_____

3:  What is the white powder suspect #2 left on the floor?
   _____Salt_____

4: Who is suspect #2? _____Joe R. Bad_____

5: Who is suspect # 3?  _____I. M. Nogood__

6:  Who is the gardener? _____I. B. Bad_____

7: Who do latent prints 1, 2, and 3 belong to?

   ___Joe R. Bad__

8: Which fingers do prints 1, 2, and 3 belong to?

   #1 __left middle___

   #2 __ left index__

   #3 ___right little finger_____

9.  Who do latent prints 4, 5, and 6 belong to?

   _____Patty Larceny_____

10. Which fingers do prints 4, 5, and 6 belong to?

   #4 ___right thumb___

   #5 ___left index___

   #6 __right middle_____

11: Who do latent prints 7, 8, and 9 belong to?

_____I. M. Nogood_____

12: Which fingers do prints 7, 8, and 9 belong to?

#7___left index____

#8 _left thumb____

#9 __ right index___

13: Which pen wrote the note found in the vehicle?

1 _____ 2_____ 3_____ 4__X____

Chemical List: TEST KEY      1993

Identify the contents of vials 1 through 17, Two (2) of the vials contain mixtures of two (2) substances.

1. Copper II Nitrate      ____14____
2. Sodium Acetate      ____9____
3. Aluminum Hydroxide      ____6____
4. Aluminum      ____13____
5. Flour/Plaster      ____7____
6. Tin      ____1____
7. Iron      ____5____
8. Sugar/Salt      ____17____
9. Sulfur      ____12____
10. Calcium Carbonate      ____16____
11. Sand      ____4____
12. Lemon Juice      ____8____
13. Household Ammonia      ____15____
14. Rubbing Alcohol      ____3____
15. Water      ____11____
16. Zinc      ____10____
17. Cornstarch      ____2____

# Event Test #2

The second year I wrote the event exam, I used the fingerprints from the first year, and changed the venue to a business and a video games store because gaming was just gaining popularity and 20 years ago microwaves were becoming a desired item. You again can identify the mountains involved if you desire to do so.

Start Time: _____   End Time:
_____

School Name: _____ Team number:
_____

Detective Names:
_____

On April 10, 1995 at 1:00 am a Video Game Store was burglarized.

Suspect #1 gained entry by climbing through the window above the back door that faces the alley way. While climbing in the window suspect #1 caught and tore his shirt on the nail, attached to the window casing. Suspect #1 proceeded to the front door and turned off the safety alarm system, let suspects #2 and #3 into the building. Suspect #1 left fingerprints on the window and inside on the alarm case cover.

Suspect #2 came into the business and climbed onto the video game. He stepped from game to game to the break room. In the breakroom suspect #1 took the microwave off the counter, knocking the condiments to the floor. Suspect #2 then took the microwave to the waiting car.

Suspect #3 went into the office and tried to open the safe to no avail. A key was found to the desk and suspect #3 took the petty cash in the drawer and the key to the coke and candy machines. There suspect #3 took the pop and candy, and delivered it to the others also.

As suspects #1, #2, and #3 came out of the store to get into the car. The owner ran up and ran after the 3 suspects, jumping into the car just as the police arrived. The police had been alerted by the safety feature in the alarm system. All 4 subjects showed identification to the police.

Suspects:

> #1. Charlie Horse
>
> #2. Shelby Coming
>
> #3. Will Steel
>
> #4. Issabelle Ringing

Police collected the following as evidence:

> #1. Material from window
>
> #2. Fingerprints from window
>
> #3. Fingerprints from alarm case cover
>
> #4. Shoe prints from the video game
>
> #5. White powder from floor
>
> #6. Fingerprints from safe and candy machines

#7. Clothing from all 4 suspects

#8. Fingerprints from all 4 suspects

#9. Vehicle was registered to I. M. Walking

#10. 4 pens found, one from each suspect to compare ink with the note found in the vehicle #11. On the alarm combination

#12. Fingerprints found on microwave


Essay and Analysis:

Please answer the following questions to solve the crime mentioned, the chemical analysis, and the essay to explain your theory and opinion of the actions of those involved. Do the chemical analysis in a safely manner, using the safety equipment at all times.


Questions: 1994 Test


1.  Does material match cloth from the shirt of suspect:

    1. _____ 2. _____ 3. _____ 4. _____

2.  Who is suspect #1?
    _____

3.  What is the white powder suspect #2 left on the floor?
    _____

4.  Who is suspect #2?
    _____

5. Who is suspect # 3
_____?

6. Who is the owner of the store?
_____

7. Who do latent prints 1, 2, and 3 belong to?
_____

8. Which fingers do prints 1, 2, and 3 belong to?

   #1 _____ #2 _____#3_____

9. Who do latent prints 4, 5, and 6 belong to?
_____

10. Which fingers do prints 4, 5, and 6 belong to?

   #4_____ #5 _____#6_____

11. Who do latent prints 7, 8, and 9 belong to?
_____

12. Which fingers do prints 7, 8, and 9 belong to?

   #7 _____ #8_____ #9 _____

13. Which pen wrote the note found in the vehicle?

   1. _____ 2._____ 3._____4._____

14. Which shoe print matches the latent lab print?

   1. _____ 2. _____3._____ 4._____

Chemical List: (Each of the solids, metals and liquids were placed in small containers for easy testing.)

Identify the contents of vials 1 through 17, Two (2) of the vials contain mixtures of two (2) substances.

1. Copper II Nitrate      _____

2. Sodium Acetate      _____

3. Aluminum Hydroxide      _____

4. Aluminum      _____

5. Flour/Plaster      _____

6. Tin      _____

7. Iron      _____

8. Sugar/Salt      _____

9. Sulfur      _____

10. Calcium Carbonate      _____

11. Sand      _____

12. Lemon Juice      _____

13. Household Ammonia      _____

14. Rubbing Alcohol      _____

15. Water      _____

16. Zinc      _____

17. Cornstarch      _____

Questions: 1994 Test Answer KEY

1. Does material match cloth from the shirt of suspect:

      1._____ 2. _____ 3. __X_____ 4. _____

1.   Who is suspect #1?

      ____Shelby Coming_____

2.   What is the white powder suspect #2 left on the floor?
      ____Sugar_____

3.   Who is suspect #2?

4.   _____Charlie Horse_____

5.   Who is suspect # 3?

      _____Will Steel_____

6.   Who is the owner of the store?

      ____Issabella Ringing_____

7.   Who do latent prints 1, 2, and 3 belong to?

      _____Charlie Horse_____

8.   Which fingers do prints 1, 2, and 3 belong to?

      #1 ___ left middle___

      #2 __left index_____

      #3___ right little finger____

9.   Who do latent prints 4, 5, and 6 belong to?

      _____Shelby Coming_____

10. Which fingers do prints 4, 5, and 6 belong to?

#4___ right thumb____

#5 ____left index_____

#6____ right middle_____

11. Who do latent prints 7, 8, and 9 belong to?

_____Will Steel_____

12. Which fingers do prints 7, 8, and 9 belong to?

#7 ___ left index____

#8___ left thumb____

#9 ___ right index____

13. Which pen wrote the note found in the vehicle?

1._____ 2._____ 3._____4.__X____

14. Which shoe print matches the latent lab print?

1._____ 2. __X____3._____ 4._____

Chemical List:     TEST Answer KEY     1994

Identify the contents of vials 1 through 17, Two (2) of the vials contain mixtures of two (2) substances.

1. Copper II Nitrate _____14_____

2. Sodium Acetate _____9_____

3. Aluminum Hydroxide _____6_____

4. Aluminum _____13_____

5. Flour/Plaster _____7_____

6. Tin _____1_____

7. Iron _____5_____

8. Sugar/Salt _____17_____

9. Sulfur _____12_____

10. Calcium Carbonate _____16_____

11. Sand _____4_____

12. Lemon Juice _____8_____

13. Household Ammonia _____15_____

14. Rubbing Alcohol _____3_____

15. Water _____11_____

16. Zinc _____10_____

17. Cornstarch _____2_____

# Event Test #3

The third and final year I was in charge of the testing, the class had become so popular that the students requested a desire to include consequences built into the scenario.  This scenario I placed in a school setting; and the consequences included what the appropriate responses should be from the school, and the families, as the 'crime' affected not only themselves, but the whole class and teacher's time.  It included the groups found within a school setting, those with goals working after school in businesses; parent's expectations; sports requirements for participation; and those who are not as socially included as they should be as they are often preyed upon and intimidated for the use of their gifts and talents to the benefit of others.


Start Time: _____

 End Time: _____

School Name: _____

Team number: _____

Detective Names:

_____


On October 15, 1995 the Math Department window was broken into, in the Franklin Public School.  All the grades on the computer were erased.  This was the night before the teacher was going to post the grades for the first trimester. The suspect broke a small section in the window and turned the inside latch.  In entering the suspect kicked over a set of flower pots the students were observing for measurement purposes.  This left a powder on the floor, and a shoe print in the dirt.  There were no large pieces of glass inside the room, only small fragments. There were no prints to be found on the computer keyboard, but the computer was not turned off. This is how the teacher observed the room the next morning. In searching the room further, one button was found, and a piece of fabric was hooked to the computer chair. The chair had been a problem to the teacher, and to the students in the past. The police were called to the scene of the crime. They were able to obtain fingerprints, and shoeprints and the powder. They also took the material, and the button for examination. Outside the window they found two sets of prints, but no glass. In searching further they found glass in the trash and could obtain 3 prints. There were shoe prints found near the trash, and another set of prints leading away from the scene of the crime. They were both going in opposite directions. The police found along one of the foot prints a note partially torn in small pieces that said, "I need help, meet me at". The police obtained the evidence before school opened, and returned to the Lab Department. You are now at the Lab and it is your job to solve this case, hopefully before innocent suspects all fail math.

Evidence found:

Of this evidence please put a checkmark by those where are only circumstantial and not hard evidence.

> #1. Broken pots, and residue powder on the Floor

> #2. Fingerprints on the window and the latch

> #3. Shoe prints inside on the floor

#4. Shoe print outside the window

#5. Shoe print outside by the trash container

#6. Button

#6. Shirt material

#7. Note in the walkway


Suspects:

#1. Will Walking: If he does not pass math, his father will not let him take drivers training. His father thinks he will not have the necessary ability for care for the car, the insurance and the necessary record keeping ability for his job in the garden shop. His shirt button is missing, and his shirt has a small rip in the back. He has a small cut on his hand that he said he got at work when he dropped some pots.

#2. Penny Stretcher: She hates math, and loves the languages. If she does not pass math, her mother will not let her skip the first 3 days of the next trimester, to attend a conference with her on the value of writing. Penny plans to take the workshop on Hyperboles, Euphemisms, and Infinitives. She has planned this trip for 3 months, and really desires to pass. She has 1 broken fingernail, and a small cut on 2 of her fingers. She said she got the cut cleaning up someone else's mess. Later in the day, at lunch, she was heard saying, "Boys are so inept, and naïve."

#3. Joe Slam: If he does not get a good grade in math, he will be cut from the basketball team. He has been on restriction for his grades in the past. He is a popular player, but can't afford the time to study. He has been caught in the past cheating by talking girls into helping him with his homework. He then lets them do the work, and he takes the answers.  He has a sprained ankle, and is getting a lot of sympathy from the girls.

#4. Harold D. Peabody:  He is a computer whiz, but is quiet and doesn't have many friends.  He is quite easily intimidated and has a small frame that has received bruises in the past by bullies.  He has tried to please the popular crowd in the past to his sorrow.  He has asthma, and is allergic to wool.  His eyes are swollen almost closed, he says he fell into some bushes.

Work Sheet:

Will:    Shirt: _____

         Fingerprints: _____

         Shoe prints: _____

         Pen: _____

Penny:  Shirt: _____

         Fingerprints: _____

         Shoe prints: _____

         Pen: _____

Joe:     Shirt: _____

         Fingerprints: _____

         Shoe prints: _____

         Pen: _____

Harold:  Shirt: _____

Fingerprints: _____

Shoe prints: _____

Pen: _____

Questions:

1. What is the residue from the floor?
   _____plaster_____

2. Whose fingerprints were on the window and latch?
   _____Joe_____

3. Which fingers do prints #1, and #2 match?

   #1. ____left thumb_____

   #2. ____left index_____

4. Whose button was found?
   _____Will's_____

5. Whose shirt did the material from the chair belong to?
   _____Will's_____

6. Whose prints were on the glass in the trash?
   _____Penny's_____

7. Which fingers do prints #3, #4, and #5 match?

   #3._____left index_____

   #4. _____right thumb_____

   #5._____middle right_____

8. Who broke into the Math Department?
   _____Joe_____

9. Who was outside the window?

    _____Penny_____

10. Who were the innocent suspects?

    _____Harold and Will_____

Punishment for the Offenders:

What should be the appropriate punishment for the offenders, check with are reasonable?

From the School:

    _____ Detention

    _____Suspension for life

    _____Failing grade for cheating

    _____Home Schooling

From the Courts:

_____Vandalism $10,000 fine

_____Death Penalty

_____Community Service

_____Probation

_____Acquittal

_____State Detention Center

From the Parents: write an essay on what the parents of those involved should do to help their children understand the consequences of their actions.

Chemical List:

Identify the contents of vials 1 through 17, Two (2) of the vials contain mixtures of two (2) substances.

1. Copper II Nitrate          _____

2. Sodium Acetate             _____

3. Aluminum Hydroxide         _____

4. Aluminum                   _____

5. Flour                      _____

6. Soda                       _____

7. Plaster                    _____

8. Tin                        _____

9. Iron                       _____

10. Sugar                     _____

11. Sulfur                    _____

12. Salt                      _____

13. Calcium Carbonate         _____

14. Sand                      _____

15. Lemon Juice               _____

16. Household Ammonia         _____

17. Rubbing Alcohol           _____

18. Water                     _____

19. Zinc                      _____

20. Cornstarch                _____

21. Sand and Plaster          _____

22.  Salt and Flour          _____



To give this test you will need:

    4 shoe prints;

    4 shirts;

    4 sets of fingerprints;

    4 pens

    1 of the shirts need to have a small rip, and the
        material as evidence

    1 wool shirt missing a button

    Plaster

    Note in trash to test the ink pens

    Photocopies of the 4 sets of prints

    Fingerprints

    Have fun!

## Conclusion:

My choices did affect my life, and the lives of my family. Their handprints and footprints could and can be seen over the choices I have made and the actions I have taken, some good and some bad,  some fun and some difficult to take. My choices did affect all 7 of my mountains.

The Religious Mountain –many from my family and local community, said I would "ruin my children" and they thought I could better "let my light shine in a dark place – public school". They thought I was "missing the mark" and I became an outcast in the religious community. They were later proud of my children and affirmed them and expressed it, but seldom to me.  But they did not agree with God's purpose and plan nor the path He had set me on.

My Family Mountain –in-laws thought I was becoming an "out-law" because some of them were in the teaching profession, but they would wait and see and hold their judgment until a later time.

The Education Mountain –wanted to take me to court because I was taking money from their state funds that they get per child.  Even though I had gone through 3 years to become a teacher in college and changed my mind, they believed I had no business teaching my own children, let alone teaching others.  Some felt I was taking awards from the school system as my children succeeded. Some said, that they were the authorities and I need to submit to their authority. Once they desired weekly surprise visits, and that we put in another bathroom for girls and boys in our home. In the end the Superintendent congratulated me on a job well done, as he went to a homeschool graduation of one of my children.

The Government Mountain – said "go for it, everything you do will be fine with us, just pay your taxes". We did not receive any tax breaks or vouchers for home education.

The Business Mountain—said "pay for it yourself". You could have made more money with us, instead of investing it in your children. You are now considered a "non-productive member of society".  We paid for all the books and resources needed to get the job well done. Each of our children, helped in the construction business, to learn skills needed for their own lives later.

The Arts and Entertainment Mountain –wanted to watch the outcome as if a "game" of right and wrong, good or evil was being played.

The Finance Mountain –yes, there is a cost to any decision, and a bottom line, there are assets and liabilities' in anything you do and every dollar you spend.  Some we pay for ourselves in many different ways, but our investments into the lives of our children and the next generation are worth every penny, even if it is blood, sweat and tears.

As to the senses, I could sense the reactions around me. Some negative and some positive; some left a bitter-sweet taste; some reactions violent as acid put onto baking soda. Some reactions were not worth mentioning as some would bless and some curse. But even now I have been putting in a mystery even into this scenario.  I do not regret my decisions to homeschool my children or to give as I have given.

Some would argue why bring God into it. Sir Isaac Newton said in 1713 in his book *Philosophiae Naturalis Princia Mathematica 2nd Edition*: "It is confessed that the Supreme God exists necessarily, and by the same Necessity

he is always and everywhere.  Whence also he is wholly
similar, all Eye, all Ear, all Brain, all Arm, all the Power of
perceiving Understanding and Acting.  But after a manner
not at all corporeal, after a manner not like that of Men,
after a manner wholly to us unknown As a blind man has no
notion of Colors, so neither have we any notion of the
manner how the most Wise God perceives and understands
all things.  He is wholly destitute of all Body and all Bodily
shape and therefore cannot be seen, heard, nor touched, nor
ought to be worshiped under the representation of anything
corporeal. We have ideas of his attributes, but we know not
at all what is the substance of anything whatever.  We see
only figures and colors of Bodies, we hear only sounds, we
touch only the outward surfaces, we smell only odors and
taste tastes, but we know not by any sense or reflex act the
inward substances and much less have we any notion of the
substance  of God.  We know him only by his properties and
attributes and by the most wise and excellent structure of
things, and by final causes, but we adore and worship him on
account of Dominion.  For, God without Dominion,
Providence and Final Causes is nothing else but Fate and
Nature."

So yes, I see God as my Creator, The Lord Jesus as my
Redeemer, and The Gracious Holy Spirit as my teacher and
my guide as I walk through my mountains and valleys. And I
am sure fate is in my own hands and my own decisions; or is
it purpose and destiny; or is it just a mystery of a great
desire, a great love for things past; or a desire to impart ideas
to the next generation.  Isaiah says, "If we forsake the Lord
and forget His holy mountain, we set a table for fortune and
fill cups with mixed wine or destiny....we then fall under

judgment because we chose things in which He did not

delight." I sincerely desire to when the time comes the Lord will say "Well done good and faithful servant."

Some would argue that I am stupid, those are not the mountains. But I said those are "my mountains". Some would say Business and Finance are one and the same, because everyone works and they gain finances and have to manage them accordingly for better or for worse.

I have said that "these are my mountains", and now I am climbing another one, the Mountain of Media and Technology. Time will tell if I am successful in climbing, or if there will be ledges to cross, drop offs to encounter, obstacles to find my footing. There will be those who will be for me and those against me, but the mountain can be climbed. So I have begun to do so. There are rivers to cross – sink or swim. So here I am stepping out to begin the climb. As before, I tell myself...."Be strong and of good courage, the

Lord is with you wherever you go", "It might not be easy but it is possible." "You can do this." "You can do all things through Christ who strengthens you."   So off I go.

Since I have written this a few years ago to explain my journey through my mountains; I have been reclaiming the mountains of the past; substitute teaching; regaining friendships and fellowships within the church, family and community.  So even if things in the past were difficult, and sometimes stressed, does not mean all is lost, or cannot be recovered in any form. So even if life has been difficult at times, remember the good, and leave a blessing for others to follow. The culture needs the mothers and fathers to show the way and prepare them to succeed in what they may encounter.  They are looking for those who will stand with them and help them to understand how they may impact their families, communities, and nation with their own gifts, talents and abilities.

Blessings!  May you prosper and be in good health having the strength to conquer and succeed in whatever mountains you desire to pursue as you move forward in your own life.

And may the event sheets inspire and give you ideas to write your own; how the material world around us can help us understand reactions, and we can become the next go to person for those who have a mystery they cannot solve. We can encourage them to step forward and begin their own climb to success

## References:

Science Catches the Criminal by Wyatt Blassingame

Simple Science Experiments by Hans Jurgen Press

Taste, Touch and Smell by Irving and Ruth Adler

7 Mountains: Lance Wallnau at:  lancewallnau.com

New American Standard Bible

How to become a Christian: 4 spiritual laws.com

www.ingramcontent.com/pod-product-compliance
Lightning Source LLC
Chambersburg PA
CBHW020711180526
45163CB00008B/3038